Fragments
Of
Desire

Verses Of Love, Temptation and Heartbreak

ANUSHIKHA KANCHAN

*For those who have tasted the sweetness of love,
danced at the edge of temptation's flame, and endured
the bitterness of heartbreak.*

There is this boy,

Who looks like poetry,

His aura feels like stardust in a perfect storm,

Thousands of galaxies dancing in his beautiful
brown eyes,

And each glance sets my heart to a love song's sigh.

On the day our eyes first met,

Within my heart, a space defined,

Angels danced, as he walked past me, glances stolen,

Flowers bloomed in the oddest part of my heart,

Each petal whispered the story of our fateful meeting, forevermore.

I like him,

And he has no clue about it,

I often find myself looking at him,

Looking for love in his eyes, a silent dream,

Hoping for a chance, our own forever,

His gaze is a whisper of possibility's dance,

And in those moments, I allow butterflies to take flight in my stomach,

Surrendering to chance for love, burning bright.

Within my heart, a symphony unfolds,

Singing his name in a poetic cadence,

In dreams, I paint his essence,

Bringing him to life through my poetry,

Wildly unsure of the aftermath,

Yet, the enigma of him is where I don't want to draw
any lines

Just hearing his name, a smile blooms,

Amidst crowds I find myself looking for his brown
eyes, a tapestry of warmth, rich and bold,

Like earth kissed by sunlight,

From deepest mahogany to honeyed wine,

Stirring emotions beyond,

Each glance, a dance of love,

In those eyes, I fall, endlessly.

Most ardently,

I want to call him mine,

I want him to know how his laughter ignites tiny
sparks on my skin,

I want him to know to love him is effortless, yet
daunting,

But most ardently,

I want him to be my forever, intertwined.

I steal a glance,

Often pondering as to what goes on his mind,

His eyes filled with mysticism, and those eyes of his can burn me setting every fiber of my soul on the conflagration,

and I would love him unabashedly even after becoming a minute part of the cosmos.

I met him,

And with every word we shared nostalgia bloomed,

His passions, and his smile, in my heart swirled,

Each spark, each glance, a sweet perfume,

Dazed by the effortless flow,

No forced affection, in our gaze it showed the
moment he walked in, love's embrace overflowed.

I held his hand,

To feel a little equanimity when all I was feeling was chaotic,

I looked into his eyes,

Only to feel at home when all I was feeling was homesickness,

I hid in his embrace,

To hear his heartbeat gave me a sense of bliss,

I was longing for a world where I can find a little love and he became that world for me.

He has an electric soul,

He brings sunshine into my life,

He makes me believe,

Always treading lightly,

Making daisies grow even in the most desolate
fragments of my heart,

Stripping away all the melancholy.

His eyes, a blend of whiskey's amber hue, and
honey's golden glow, a spell imbue,

His gaze, a force, leaving me in disarray, with every
look, my heart began to sway, in those sunset eyes,
all worldly cares deterred,

In his eyes, I found solace, such magic he is.

I said,

"Darling,

I want to love you in different million ways, through all our versions, let our hearts entwine, our souls dance free,

Promise me, that you won't be afraid of the force of love,

Let raw emotions, like hues, collide, on passion's palette, creating magic".

His heart always radiates sunshine,

A heart so gentle yet so strong,

He grows flowers in the oddest places,

Tearing away all the worries,

His smile soothes the storms inside me,

Showering love like lilies, carrying me forward.

He reminds me of simple things,

He reminds me of the luminescence around me,

He reminds me of all the things that bring hope to
the universe,

He reminds me of love and kindness which still
exists in this world,

He reminds me of home.

Darling,

Be my golden hour,

Where time stands still looking magical

the hues of our love embracing us,

Leaving a trail of stardust,

As I write love letters to your heart at dusk.

In his eyes, I see the stars,

In his arms, peace soothes my heart's scars,

As we slow dance, our fingers intertwined,

He promises to catch me if I ever fall,

So here's my heart, my darling,

Be my always and forever.

Everything with him comes easy,

Talking for hours,

Laughing at silly things,

Even the silence is comfortable with him,

I cannot believe how he has me falling in love with him,

Constantly,

Consistently,

And continually.

I found all I have ever dared to dream,

A love so rare, like a whispered theme,

He stirred something in me so perfectly,

My heart's no longer mine, it's his to begin.

How strange it is that I dream of you,

Placing my fingers through your hair,

Blurring all the lines,

Tempting me to have my way with you,

Devouring every inch, leaving me elated on you,

Until the warmth of the morning seeps in.

He's a gentle breeze, a soft caress, easy to love, yet a
thrilling quest,

He listens to my silence and understands it,

And then he kisses me with all that he has,

Different from the rest, my heart believes,

Because whenever he looks at me it's magic.

I often see us as the thistle queen and silent king,

I often see us as Persephone and Hades,

I often see us as a beautiful disaster,

Loving despite the scars we both carry in our hearts,

Piercing through the tangles of doubts,

I often see us as saints who will be happy ever after.

Unravel me in the silk sheets,

Hug every inch of my skin,

Make my heart and soul drip in the sweetest honey-like bliss, whisper stardust on my lips,

Just once make me feel you are real, and not just a beautiful dream.

Wrap me in your poetry,

Dress my skin with your words,

Make love to every corner of my lips,

Let me breathe the butterflies,

Lift me with your passion and warmth,

And let me be your forever favorite song.

It's strange how his touch ignites a million sparks on
my skin,

How a single look from him melts all my worries
thin,

A mere thought of him makes my heart swell with
love,

Oddly, his presence always feels like fresh flowers
and an old love.

I revel in painting my feelings on his lips,

Crave to see him tremble, my masterpiece's eclipse,

I want us to keep talking all night, rhythmically,

speak love to him and spark his soul,

and for once I crave to be passionate, wild,
prejudiced,

as I watch him consume every drop of my love
unabashed.

His lips feel like whiskey,

His eyes looking into mine, consigning to oblivion,

Making me question the language I am fluent in,

He makes me feel like I am ruined for any other
love ambles into my life, making everything and
everyone else inconsequential to their existence.

He whispers stars against my lips,

Creating new galaxies on my skin,

Singing an ode, as euphoria dances between my lips.

I ache for you, solely mine,

Touch me, ignite my senses, intertwine,

Speak to my body, surrender to wilderness
unrestrained,

Unleash your desires, let them abscond,

Surrender, let passion's flames, untamed, be
unchained,

Watch my body sing, in ecstasy, is pure and right.

It's a weird, loud and crazy world out there,

Yet your heart whispers in a different language, that my love is poetry,

Undefined, Unspoken, a transcendent dream.

Love is an art,

And he is like Van Gogh,

Painting starry nights on my skin,

Taking his time with all the brush strokes on canvas
filling with the color of love.

Caramel kisses,

Blankets me with warmth sleeping over every inch
of my flesh,

Leaving butterflies in every part of my ribcage,

Fill my aura deep with the hues of your love,

Let me be the favorite note you play tonight.

Leave me gasping,

Leave me breathless,

Spin your touch like a dance, against the wall, our passions clasp, hands upon my neck,

Wild within, show me the Milky Way, as we reach higher and higher until we see the stars.

I crave to lie beside you, your touch, a gentle sigh,

Through tangled locks, your fingers glide, whisper desires making me fly,

I ache for you,

I yearn for you,

Just like the ocean craves the shore, you see

Perhaps if you allow me to embrace your lips, I could
show you how desperately my heart yearns for you,

how desperately I want to melt under your ardent
trace,

Desperate to make you my muse,

Perhaps if you allow me, I could show you how I feel
explosions, and how hard it is for me to string my
thoughts together,

Perhaps if you allow me I could show you how it
feels to be under your spell.

I want to imprint on your soul,

So you can witness my love for you,

So you can live forever in my art not just as my lover
but as my eternal muse.

Map every corner of my body,

Trace each curve, every depth, seeing until my body is fervent, blanket me with you kisses,

Paint me with your love, each hue reminisces, as euphoria dances in effortlessly between our blisses,

Unwrap ourselves, all the layers of guise, for in your arms, amidst chaos and demise, I'd choose you, my love, come what may.

Not my first love, nor yours. The first kiss eluded our embrace. Yet, in my longing for love's respite, you came in like my savior —a touch imperfect, almost mine. In that fleeting dance, I named you mine, a whispered 'forever' to the most beautiful 'almost'.

I will never be whole again,

Scars will forever dwell in my heart, a reminder of
the cost of loving you.

I was mistaken to think you could heal my soul;
instead, you left black holes in it.

I thought I was clear with my words and my actions
when I told him what he meant to me,

Yet, he chose to be oblivious, unraveling the threads
that bound us,

while I was busy building sandcastles, he was busy
stomping on each one of them,

While I was busy picking the prettiest blooms of
affection, he was busy setting the garden ablaze,

And before I could save my love, I was left with a
heart adorned with thorns.

Oh, to unbind my heart from you,

Untangle your essence from the depths of the soul's hue,

Forget all the dreams, erase your shadow,

Forget the whispered promises, untrue,

I yearn to affirm my worth, to shatter the illusion,

To sever the ties, to find a better for me,

If only I could silence the relentless chatter of my love for you.

I spun illusions,

Fooled by wounds, mistook for light,

Dreamed you'd be the one to water the flowers that
were dead,

Fooled myself, thinking this chance would set my
heart's flight.

I wish to hate him,

To wash away the memories etched within my heart
like thorns,

To purge the love once treasured, now shrouded in
pain's cruel Art,

I wish to forget all the Hurt, tearing my fragile part,

To sever the tether of hope, weary of it's ceaseless
Chase,

I wish to let go of him and set myself free.

Betrayed,

Confused,

Enraged,

Here I bleed in shadow's embrace,

Lost in a crimson tide's gentle grace,

Wondering how those hands, who once promised
me to catch if I ever fell

Be the ones to tear me apart?

You know what baffles me is that we could have
made it work only if you had loved me enough,

Only if you had told me that day on the call, that I
love you and we could get through this but instead
you chose to shatter, leaving fragments to find,

Ensuring my heart's ruin,

That day when I said "We need to talk", with dread,

You ensured heartbreak, leaving no peace behind
severing the love's thread.

My heart's been fractured, laid bare,

By a man I deemed to be mature,

I craved tender love, in his embrace,

Dreamed for honesty in his gaze,

Instead found manipulation, where love hides,

He filled me with hope, only to witness my descent,

Hurting me like a boy, not the man he was meant.

We were a puzzle, perfectly aligned,

Now, shattered fragments, our essence unbound,
each piece fractured, lost in twilight,

How do we mend what's taken flight?

He left as a sunset fading,

Leaving bittersweet memories of our end,

Now heavy is my heart,

Clinging to a phantom of what was never mine,

Deluded me into believing,

Now I bleed on paper so my heart can heal from his dearth.

Is it bad that two years later,

I'm still writing you love letters.

Is it bad,

That I clutch memories of you right?

Is it bad that,

I'm still clinging to the version of yours that was in
love with me,

Is it bad that,

I still listen to your favorite songs just to remember
you by them,

Is it bad that,

Every time I see you I hope that we can go back to
how it was.

Your memory lingers like a ghost in the air,

Haunting every part of my being,

The moments we shared,

Dreams we bared,

Now leave me wrapped in despair,

The sweet nothings you once told me,

Bring an ache in my heart when I think of you,

Your echoes will forever cling, reminding me of the possibilities.

Homesickness is a wretched ache,

They warned, "Don't make him your home,"

Wearing rose-coloured glasses, believed he'd fill
every void and love me whole, alone,

Alas, I was wrong,

He burned everything beautiful for me,

Leaving only ashes where our love once belonged.

I wonder, when did it begin,

When did your feelings for me grow thin?

When was the moment your "I love you" turned
untrue,

When did you realize I was not the girl you dreamt
of,

When did you decide that you didn't want to call me
yours,

When did you replace me without a second's
shame?

I wonder, when did you truly give your heart away,

Or was it never mine, even for a day?

At night my sorrows knock on the door,

Lurking in the darkest corner of my head,

Waiting for me to let them in,

Mourning the things I had wished to forget,

Colouring my dreams with the midnight color takes
away the faint light which I let in every time.

You were my every wonder,

My lyrical muse,

My favorite flowers,

My favorite song,

My golden light,

A hint of magic,

But alas, I was none of this for you,

A shame, you couldn't love me as I loved you,

For me you were all I ever dreamed, wished, and desired.

One day,

I will stop falling in love with you,

One day,

My heart's fractured pieces will renew,

One

I'll no longer drown in thoughts of you,

One day,

I will erase you from within.

At the hush time of night, I often rip out the pages
of painful memories and burn them in the winter
fire, seeing them dancing along with the flames,
absorbing all the pain within,

Bringing relief, giving a whimsical rhythm to me.

I go home with a heavy heart,

With no one to rest my head on shoulder,

With no one to wipe my tears, so I'm learning to
keep them inside, learning to keep myself in control,
knowing no one will come to my rescue, knowing
that I'm going through the pain which can be
overwhelming but also knowing that I'm going to
get better.

Things can't always be perfect,

Just as stars don't shine all the time,

Some day you will find someone who will sing the same rhythm as you,

Some day you will not be someone's maybe,

There will be someone who will chose you every time in their every lifetime.

I am trying to unlearn you,

Rewriting my story, tearing out the pages where you
dwelled,

The ache is bittersweet as I surrender each day,

Recalling who I was before you.

I had forgotten that I can grow flowers too,

I forgot the magic within me,

I forgot I can love myself too,

More than anyone ever would,

So now, I will watch myself bloom into the prettiest
flower,

And fill my aura with poetry, stardust, and love.

I realized I must end the war within,

Banish the shadows of self-doubt,

Reclaim myself by nurturing the cuts and bruises left
by him,

And love myself a little more,

Everyday.

Each time I look at myself in the mirror,

Reveals a canvas of love, filling the fractures with love,

The love I gave myself,

As voids blossoming into daffodils,

Telling the journey of a girl who once shattered, is now healing and is resilient to find new ways to love.

I once thought happiness lies only with the person
you love,

But now I know a deeper truth,

True happiness resides in self-love,

Embracing every shade and hue,

In the mosaic of our being, both good and bad.

A reminder to ourselves,

It's okay if you don't have it all figured out,

It's okay to test the waters of uncertainties,

Beauty is in accepting the unfolding journey,

Where each experience adorns the canvas of the
soul.

The realization of anything often arrives with strife,

Not all truths are gentle,

To confront the void where a relationship once
dwelled,

To linger in the shadow of a bond devoid of
understanding,

To realize not being in love with yourself,

To realize that you're in a different place than
everyone else and that your life will never be as you
imagined.

To witness the future's unforeseen turns,

Each realization is a tapestry woven with hues both
radiant and somber.

I am still learning the difference between love and
love's illusion,

But should love to knock on my door again, I'll open
my heart's door embracing it without debate.

Because love is love,

And just like after every storm, there is light dancing
across the sky,

After heartbreak comes the blooming,

Where the heart, once fractured, finds its beat,

And dances again to the rhythm of its own song.

ACKNOWLEDGEMENTS

First and foremost, big thanks to my readers for picking up this book. I love and appreciate every single one of you.

Secondly, thank you much to my family for always believing in me and being my support system.

And how can I forget my friend Akash, who has supported me through the journey of this book, instilling the very idea of this book and helped me develop the theme of this book. This book definitely wouldn't be out there without you.

www.ingramcontent.com/pod-product-compliance
Lightning Source LLC
Chambersburg PA
CBHW021133130726
47988CB00003B/1281